KNOT KNOW-HOW

Steve Judkins and Tim Davison

FERNHURST|BOOKS

Cover design by Simon Balley
Design & DTP by Creative Byte
Printed in China through World Print

Contents

Introduction

Ten knots everyone should know

Tying a rope to an object (with a hitch)

Tying two ropes together (with a bend)

Loops

Stopper knots

Tools of the trade

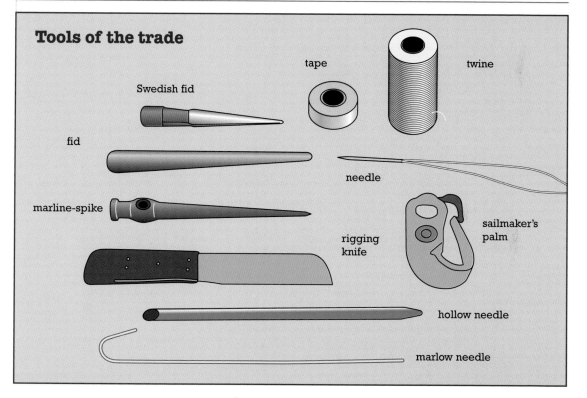

Welcome to KNOT KNOW-HOW

This book begins with the ten knots everyone should know (pages 10 to 29). It then gives lots of other knots, arranged by use (e.g. knots for tying a rope to an object). Finally, we show how to whip (stop the end of a rope unwinding), seize (sew or bind two ropes together), splice (join ropes permanently), taper and handle ropes.

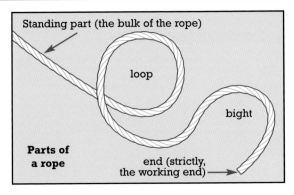

Parts of a rope

Standing part (the bulk of the rope)

loop

bight

end (strictly, the working end)

Terms

Bend	A bend joins two ropes.
Hitch	A hitch is used to attach a rope to something else, eg a post.
Splice	A splice is used to make an eye or a join without tying a knot. It works on friction. It doesn't weaken the rope as much as a knot.
Seizing	Joining ropes together by sowing or binding.
Whipping	A thin line used to stop a rope's end unlaying (unwinding).
Lay	The direction of twist of a rope's strands.
Loop	A complete turn, with a cross-over.
Bight	An incomplete loop.
Stopper knot	A knot which stops a rope being pulled through an eye.

Swedish fid	A grooved spike for splicing rope.
Coil	To twist a rope into a series of loops.
Thimble	A metal fitting put into an eye (to reduce wear).

Knot basics

A half hitch is the start of many knots.
So is a round turn.

Security

Length of tail Always work the knot tight, leaving a good tail.
Strength A knot reduces the strength of a rope. (Some knots may almost halve the strength.)

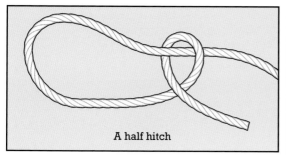

A half hitch

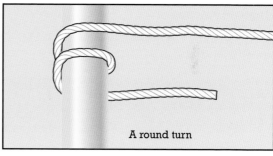

A round turn

Undoing a knot

Some knots can be capsized – eg a reef knot.
Hold the standing part and pull back the nearest end.
You can then slide the knot off the rope.

ROPES

Ropes can be made from a range of materials.
The table on page 8 lists some of these, and the
properties of the ropes.

Once the fibres have been chosen, they can be
put together in a number of ways to make a rope.
The most common ropes are *3-strand laid, 8-strand
braid* and *core-plus-cover* (see page 9).

Pre-stretched rope

The rope is stretched in the factory.

When you use it later, it won't stretch so much.

Wear

Wear causes broken fibres, or can even melt the
fibres together. To avoid this, lead ropes properly
so they don't go round sharp corners. Be careful
with core-plus-cover ropes: the strength is in the
core, but it's hidden by the cover.

Sunlight

The ultra violet in sunlight is bad for ropes.

Measuring ropes

The thickness of a rope is given as the diameter
in millimetres. (Ropes used to be measured around
the circumference, in inches.)

Choosing a rope

1. First decide on the strength you need.
2. Then decide how much stretch you need.
3. Do you want the rope to float or sink?
4. Do you need a soft rope (for easy handling?)

What ropes are made from		
Type of fibre	**Name**	**What's the rope like?**
Synthetic	Nylon	Smooth. Stretchy. Very strong.
	Polyester & Terylene	Smooth. Stretches a little. Very strong. Common. Heavy.
	Polypropylene	Can be smooth or hairy. Floats. Cheap.
	Polyethylene Plastic	Floats. Weaker than Polypropylene. Cheap.
Exotics	Spectra & Vectran	All the exotics are stronger than steel rope, size for size. All are expensive. All have very little stretch.
	Aramid	Aramid is not very good at bending and weakens when knotted. Doesn't wear well.
Natural	Manilla & Sisal	Cheap. Weaker than man-made fibre. They rot.

How is the rope made?			
Type	**Name**		**What's it like?**
3-strand laid	hawser laid		Traditional. Easy to splice
8-strand braid	multiplait		Withstands jerking: tow ropes, anchor ropes
core-plus-cover	braid on braid		Flexible. Strong. Core & soft cover (for easy handling).
core-plus-cover	braid on 3-strand		Flexible. Strong. Core & soft cover (for easy handling).

There are many alternative ways of tying the knots in this book. And most of the knots have alternative names! As you become more proficient, you will find your own preferred methods and names.

Ten knots everyone should know

Round turn & two half hitches

Use: Attaching a rope to a ring or post.

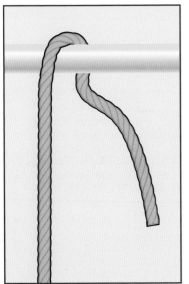

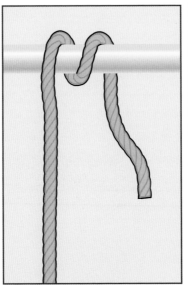

1. Pass the end round the object.

2. Take another complete turn.

3. Take the end over the standing part, around it and back through to form a half hitch.

 One of the most useful knots.

Secure, if tied correctly
and tightened up.

Good for untying under
pressure. Provided you
keep tension on the end,
the round turn will hold
the loaded rope while
you untie the half hitches.

4. Repeat, to form a second
half hitch.

5. Pull tight.

Clove hitch

Use: Attaching a rope to a ring or post.

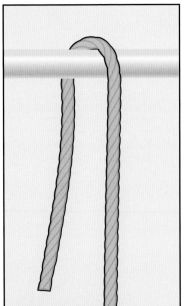 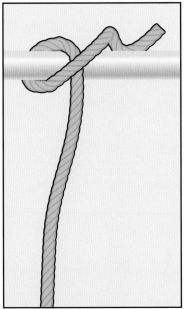 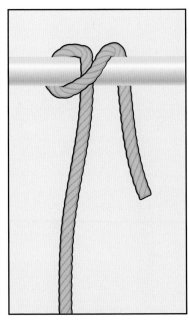

1. Pass the working end over the object...

2. ...and back over the standing part.

3. Pass the working end round the object.......

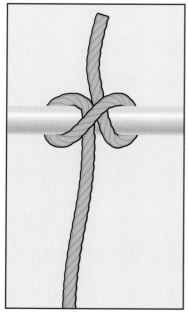

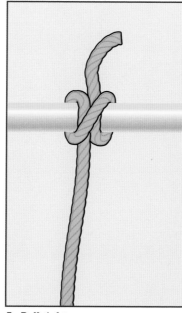

 Easy to undo.

 Must be an even pull, on both ends.

 When the pull is from one end only, the knot can slip and work loose.

4. .....and back through the loop.   **5.** Pull tight.

Figure of eight
Use: As a stopper knot. Stops the end of a rope being pulled through a hole.

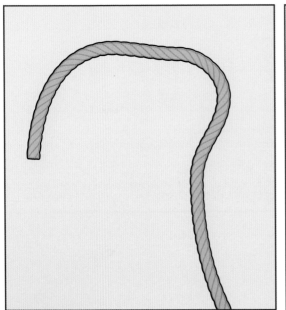

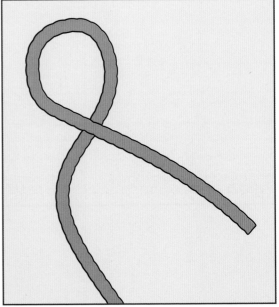

1. Make a bight.

2. Pass the end over the standing part to form a loop.

Easy to undo.

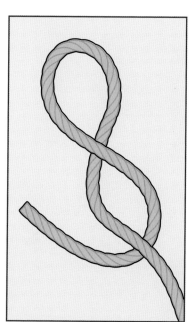

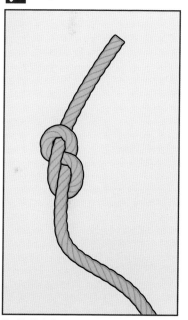

3. Pass the end under the standing part.

4. Pass the end through the top loop.

5. Pull tight.

Reef knot

Use: For tying the ends of a rope around an object, e.g. a parcel, a bandage, the neck of a sack.

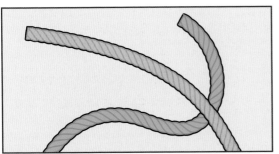

1. Keep working with the same end. Right over left.

2. And under.

5. ...and under.

6. Pull tight and check.

Note: A bow is a reef knot, with steps 4,5 & 6 made from loops.

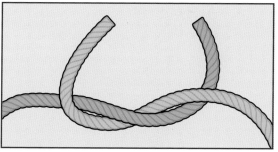

3. Carry on with the same end....

4. ...left over right...

If one end snags, or only one end is pulled, the knot can capsize and come undone.

Check you haven't tied a granny knot. This will slip.

Bowline

Use: Making a secure loop in a rope.

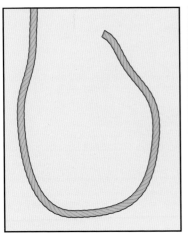 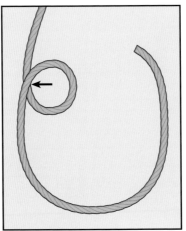 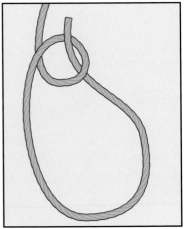

1. Form a bight of the required size.

2. Make a small loop.

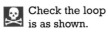 Check the loop is as shown.

3. Pass the end up through the small loop .....

4. ...under the standing part.......

5. . ...and down through the small loop.

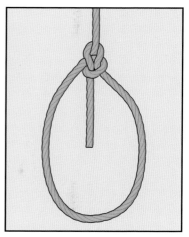

6. Pull tight and check there is a long tail.

 You need a long tail.

 Easy to undo, provided you can take the strain off the rope.

Bowline on a bight
Use: A chair or a harness.

 A secure knot. Can still be tied if there is no free end.
The working bight is the mid-part of the rope.

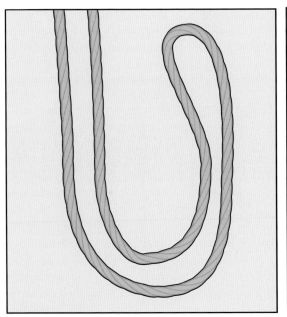

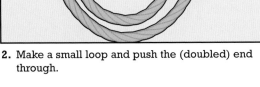

1. Form a bight.

2. Make a small loop and push the (doubled) end through.

Running bowline

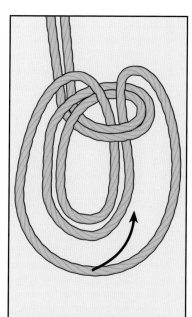

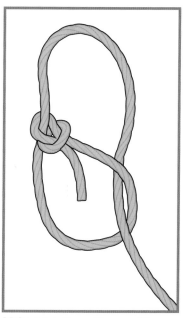

3. Open out the end and take it over the bottom of the knot.

4. Push it right under the knot.
5. Pull tight.

Make a bowline with a small loop. This loop runs on the standing part.

Sheet bend

Use: Joining two ropes of similar thickness.

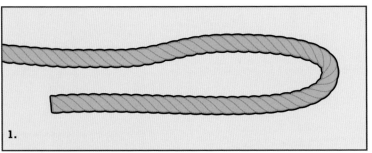

 Where the ropes are very different in thickness, use a double sheet bend.......
see pages 24-25.

1. Make a bight in the thicker rope.

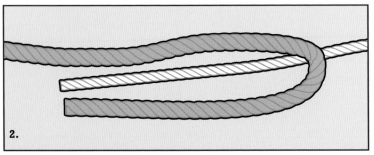

2. Pass the thinner rope through the bight .....

3. ....around and under, in the direction that will eventually leave both ends on the same side.

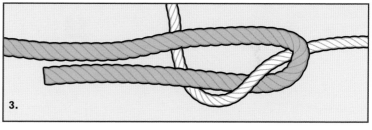

4. Pass the end of the thinner rope under its own standing part.

5. Pull tight. Double check that the loose ends are on the same side.

Double sheet bend
Use: A more secure version of the sheet bend.

 Check: Both ends must be on the same side

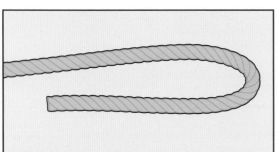

1. Make a bight in the thicker rope.

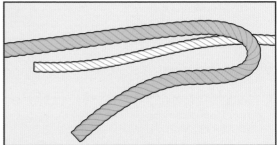

2. Pass the thinner rope through the bight.

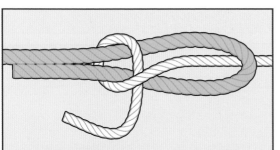

4. Complete a single sheet bend.

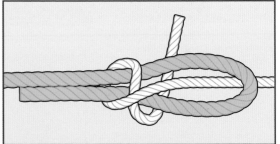

5. Pass the end under the thick rope again.

More secure than a single sheet bend.

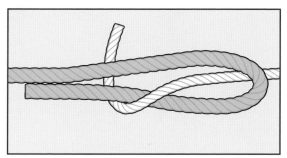

3. Pass it over the thick rope, then under both parts of it.

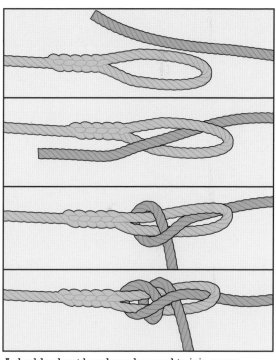

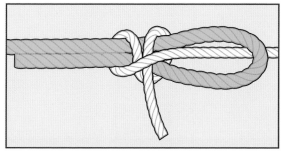

6. And under its own standing part.

A double sheet bend can be used to join a rope (of any thickness) to a rope loop (becket).

Fisherman's bend/Anchor hitch

Use: Attaching a rope to a ring (e.g. on an anchor).

This knot is related to the round turn and two half hitches, but is more secure.

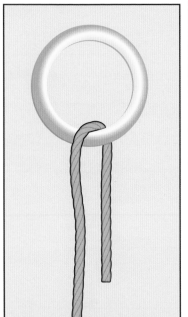

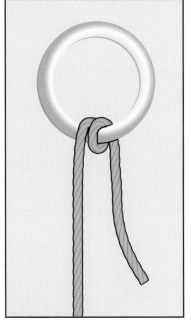

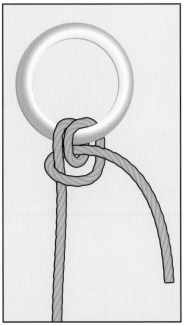

1. Pass the end through the ring.

2. Make another turn.

3. Pass the end through the turns.

Very secure.

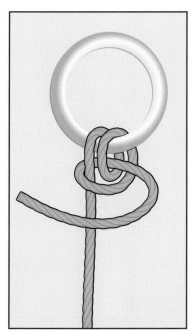

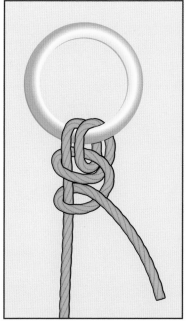

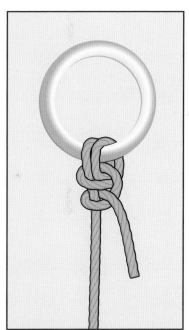

4. Then over the standing part.....

5. ...and under to make a half hitch.

6. Pull tight.

Rolling hitch

Use: To attach a line to a rod or
another rope so it grips it.
To pull a log.
To take the strain off a fouled rope.

1. Pass the end over
   the object.

2. Take it around the object
   and over itself.

3. Take it around the object and
   over itself again.

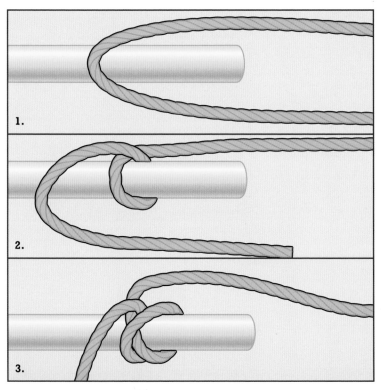

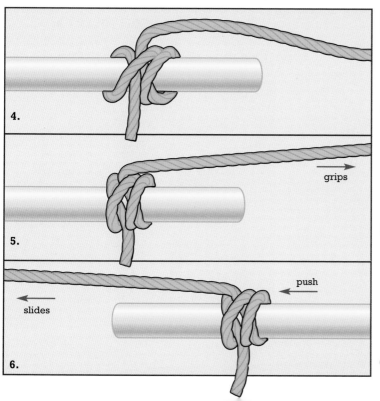

☠ The knot only grips in one direction. Plan ahead, so it will grip when you pull.

4. Take the end around the object again and back under itself.

grips →

5. The knot grips when pulled like this.

push ←

← slides

6. The knot slides when pushed the other way.

Tying a rope to an object

A **hitch** is used to tie a rope to an object.

Summary: We have already learned four hitches:

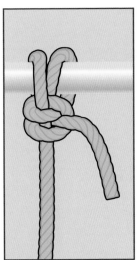

Round turn and two half hitches (pages 10-11)

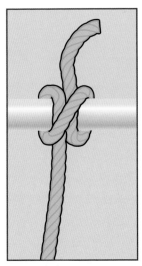

Clove hitch (pages 12-13)

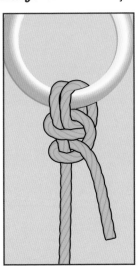

Fisherman's bend (pages 26-27)

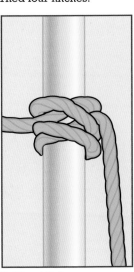

Rolling hitch (pages 28-29)

Now let's look at thirteen other hitches....

Buntline hitch

Use: A safe knot for securing a rope to a ring or post.

✅ Very secure. ☠ Not easy to undo

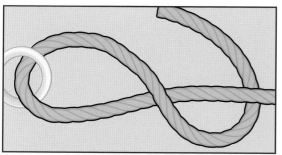

1. Pass the end through the ring and over then around the standing part.

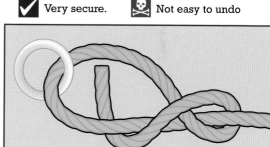

2. Back over and under the standing part.

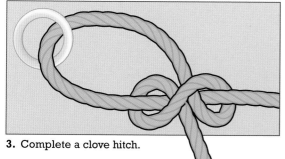

3. Complete a clove hitch.

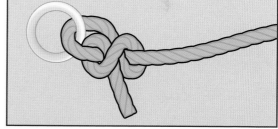

4. Pull tight.

Constrictor knot

Use: To make a temporary whipping on a frayed rope. A binding, e.g. to tie up the top of a sack.

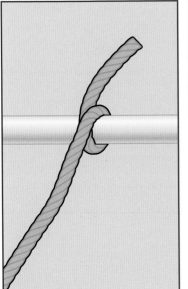

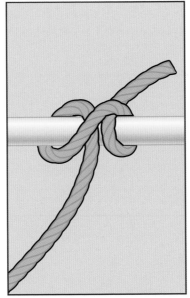

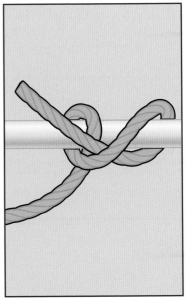

1. Take the end around the object and over itself.

2. Then around again and back under itself. You have tied a clove hitch.

3. Now take the end sideways, across the standing part.

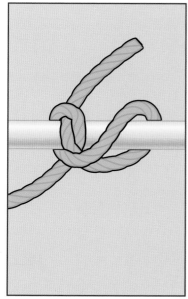

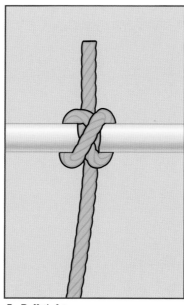

 Very hard to undo.

Very secure.

4. And down through the loop.

5. Pull tight.

Surgeon's knot

Use: For tying the ends of a rope around an object, e.g. a parcel or a bandage.

Like a reef knot with an extra twist.
The extra twist helps hold the knot while tying the second part.

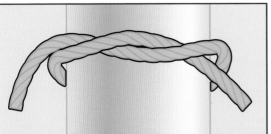

1. Right over left.

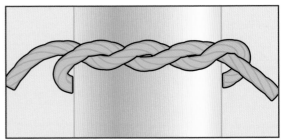

2. And again.

3. Left over right.

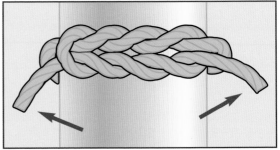

4. And again. Pull tight.

Timber hitch

Use: To lift or drag a long, round object.

 Easy to undo.

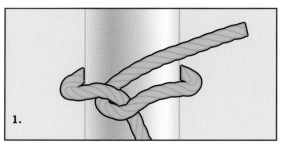

1. Put the end around the post and
   around the standing part.

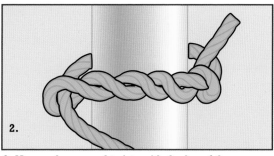

2. Now make several twists with the lay of the rope.

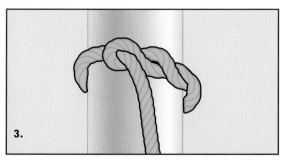

3. Draw the hitch up tight.

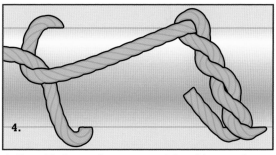

4. For added security, go around the object again.

Cow hitch

Use: To attach a rope to
a peg or rod.

Insecure, unless the free end
is locked as in figure 4.

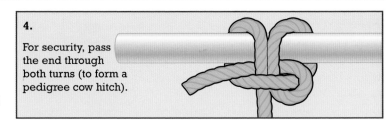

4.

For security, pass
the end through
both turns (to form a
pedigree cow hitch).

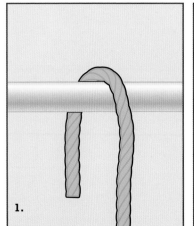

1. Take a turn around the object.

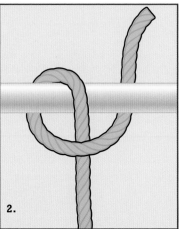

2. Then over the standing part and
under the object.

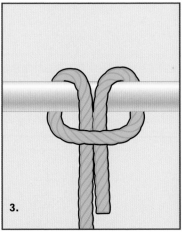

3. Round the object and back
through the loop.

Prusik knot/Cow hitch round turn

Use: To attach a rope to a pole. To help a climber climb up his safety rope. In this case he would use a small loop of rope, with a prusik knot tied in it around the safety rope. It slides up OK, but locks when he puts weight on it.

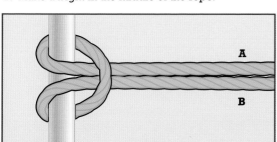

1. Make a bight in the middle of the rope.

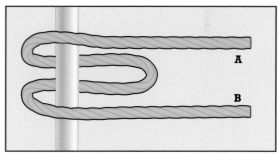

2. Put the bight around the object.

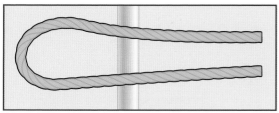

3. Put A and B through the bight.

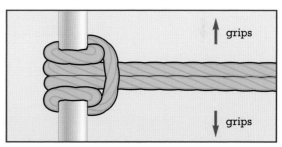

4. Take the bight around the object again and put A and B throught the bight again.

Marlinespike hitch

Use: Attaching a rod to the middle of a rope, to use as a purchase.

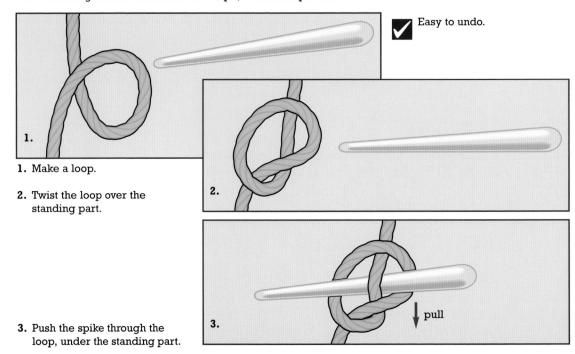

✓ Easy to undo.

1. Make a loop.

2. Twist the loop over the standing part.

3. Push the spike through the loop, under the standing part.

pull

Tautline hitch

Use: To make a loop that slides in one direction (making the loop bigger)
but locks when under strain. The knot is similar to a rolling hitch.

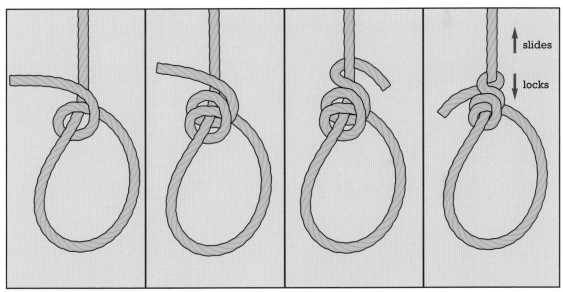

1. Take the working end over the standing part, around it and over itself.

2. Make another complete turn.

3. Take the working end under the standing part....

4. ....and back through itself.

Klemheist knot

Use: To attach a rope to a line or rod so it slides in one direction, grips in the other. Use it to climb a mast, sliding it up between 'steps'.

 Similar to a rolling hitch but has greater gripping power.

 Only grips in one direction.

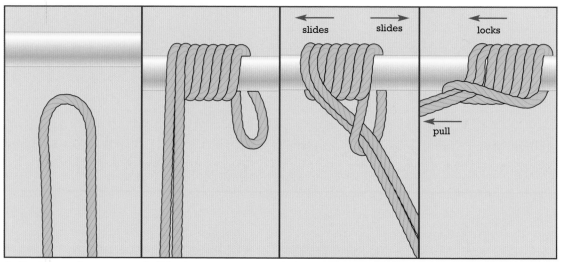

1. Double up the rope.

2. Take four turns around the object.

3. Pass the ends through the loop. While slack, the knot will slide in either direction.

4. It grips if pulled to the left.

Stunsail halyard bend
Use: Attaching a rope to a pole
or a spar.

 Excellent holding power.

 Difficult to undo.

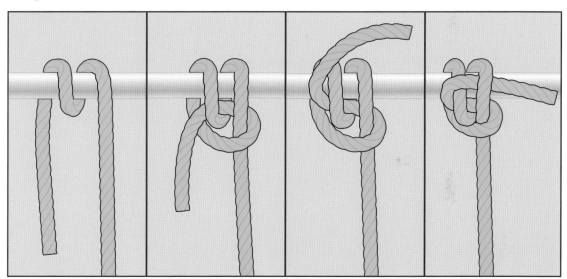

1. Make a round turn
   around the object.

2. Take the working
   end over the stand-
   ing part and through
   the round turn.

3. Take the working
   end over one of
   the turns....

4. ....and under the
   other turn.

Draw hitch

Use: Attaching a rope to an object so it can be released instantly, and from a distance. E.g. attaching a rope at the top of a mountain, then releasing it when you are at the bottom.

Keep pressure on the standing part.

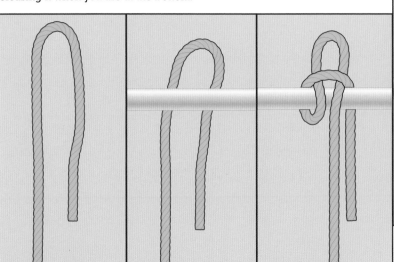

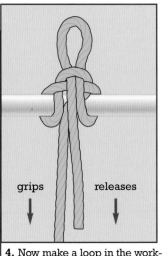

grips   releases

4. Now make a loop in the working end and pass it through the loop you made in step 3. Pull tight. The knot holds while pressure is on the standing part, but comes undone when the working end is pulled.

1. Double up the rope. The 'releasing end' has to be long enough for you to release it.

2. Pass the loop under the object.

3. Pass a loop of the standing end up through the first loop.

Magnus hitch

Use: Securing a rope to a pole with the pull at 90 degrees to the pole.

 Needs a steady pull to be secure.

This knot is like a clove hitch but with an extra turn, so is more secure.

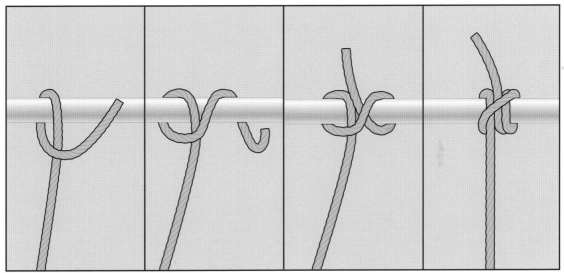

1. Pass the working end around the object.

2. Pass it around the object again....

3. ....and under itself.

4. Make a further turn.

Lighterman's hitch

Use: Attaching a rope to a rod, peg or post.

 You don't need a rope's end.

 Holds well but can be undone easily and under load.

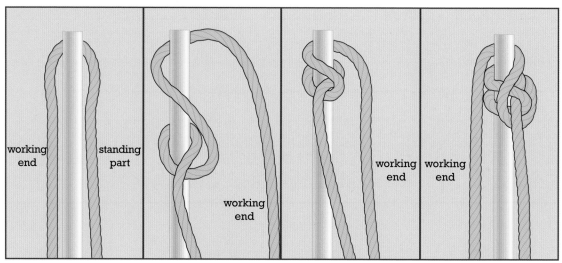

1. Put the rope round the post.

2. Take a loop of the working end under the standing part and over the post.

3. Take another loop of the working end and pass it under the standing part and around the post in the other direction.

4. Repeat as many times as you like.

Tying two ropes together

A **bend** is used to tie two ropes together.

Summary: We have already learned three bends:

A reef knot.

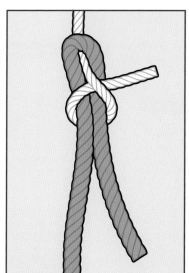

A sheet bend.

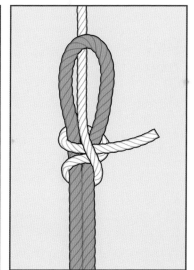

A double sheet bend.

Now let's look at five other bends....

Carrick bend
Use: For joining large ropes as well as small ones.

1. Make a loop.

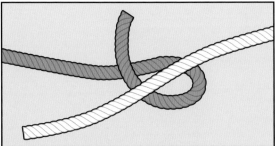

2. Pass the second rope over the loop.

5. Under.

6. Over.

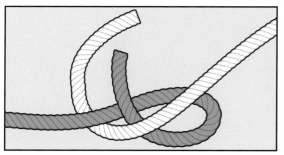

3. Under.

4. Over.

7. Under. For large ropes, complete by lashing the ends to the standing parts.

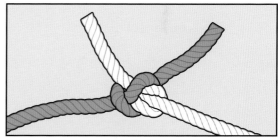

8. For small ropes, don't fix the ends. Instead pull until the knot flips, like this.

Alpine butterfly bend

Use: Joining two ropes securely.

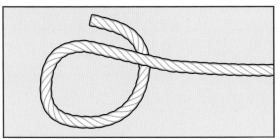

1. Make a loop with the end under the standing part.

2. Take the second rope through the loop, round, and under its standing part.

3. Take both working ends down through the centre of the knot.

4. Pull tight.

Fisherman's knot/Englishman's knot

Use: Joining two ropes securely.

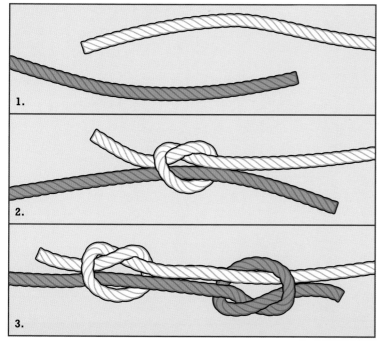

 For thin line.

 Difficult to undo.

1. Lay the ropes beside each other.

2. Take one end round the other, under and back through it's own loop.

3. Repeat with the other end. Pull tight so that the two knots slide together and lock against each other.

Zeppelin knot
Use: Joining two ropes of
similar thickness.

 A very strong knot
(used to tether airships!).

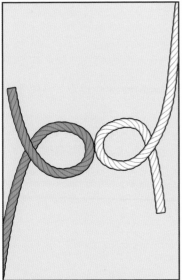

1. Make a loop in each rope like this.

2. Put the left-hand loop on top of the other one.

3. Pass the working end of the upper loop under everything and up through both loops.

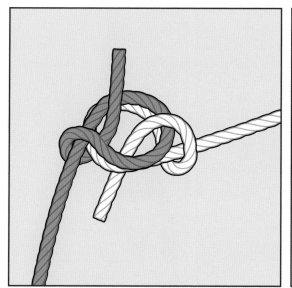

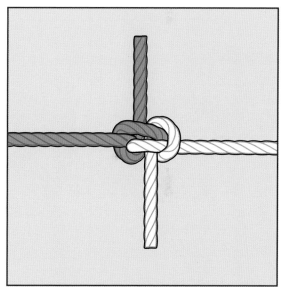

4. Pass the working end of the lower loop down through both loops.

5. Tighten the knot carefully so it falls into place.

Hunter's bend

Use: To join two ropes of similar thickness. Invented by Dr Hunter in the early 70's, this is one of the most recent knots.

 An excellent knot for modern (slippery) ropes.

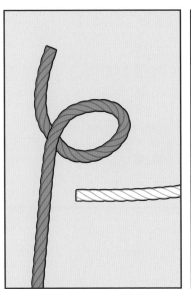

1. Make a loop.

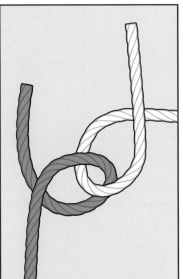

2. Pass the second rope through it and make a second loop.

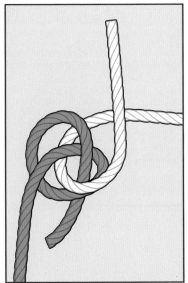

3. Take the working end of the first rope down through both loops.

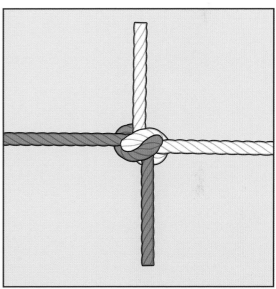

4. Take the working end of the second rope and pass it up through both loops.

5. Pull tight.

Loops

Summary: We have already learned three loops:

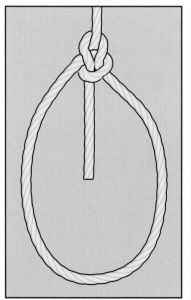

Bowline.

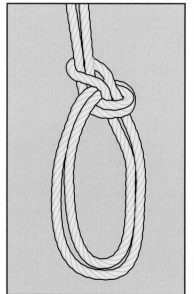

Bowline on a bight.

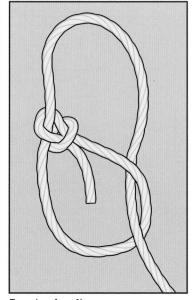

Running bowline.

Now let's look at six other loops....

Man harness knot / Artillery loop

Use: To make a handle in a rope (e.g. for pulling).

 Easy to undo.
No need to have an end.

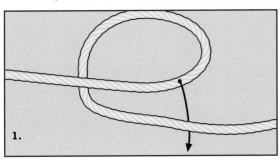

1.

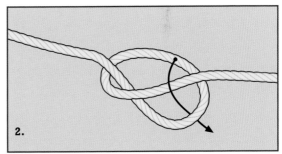

2.

1. Make a loop in the middle of the line.

2. Slide the loop behind the standing part.

3. Take the top of the loop over the standing part and under the bottom of the loop.

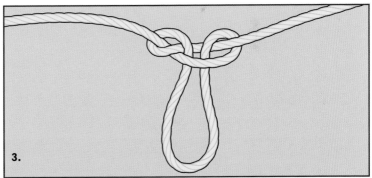

3.

Figure of eight loop

Use: Attaching a rope securely, e.g. fixing a rope to a climber's harness.

 A very secure knot.

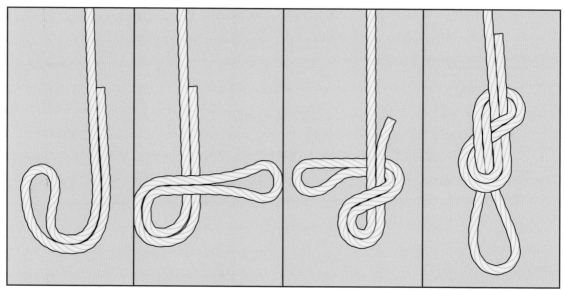

1. Double up the rope.

2. Pass the loop over the standing part....

3. ....and under it.

4. Pass the loop through the bottom of the figure of eight.

Threaded figure of eight loop
Use: Attaching a rope to a ring.

 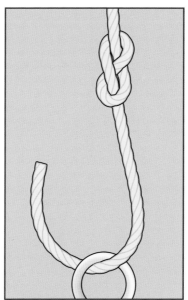 

1. Make a figure of eight knot with a long tail.

2. Pass the working end through the ring.

3. Thread the working end through the knot, following the figure of eight.

Adjustable eye/Standing eye loop
Use: To make a loop that can be undone easily.

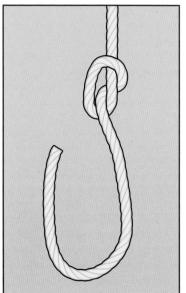

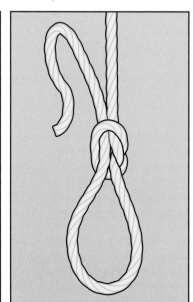

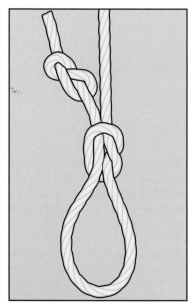

1. Make an overhand knot (see page 64).

2. Pass the working end through the overhand knot to make a loop.

3. Adjust the size of the loop. EITHER tie a figure of eight knot in the working end OR....

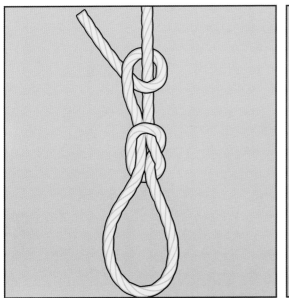

4. ....tie a half hitch around the standing part OR....

5. ....tie a half hitch under the overhand knot.
In each case, pull tight.

Angler's loop

Use: Making a very secure loop in a rope.

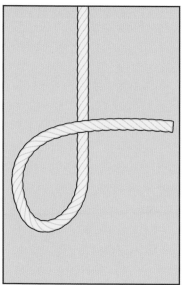

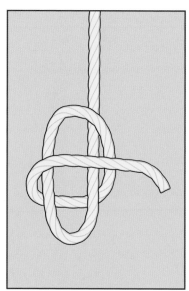

1. Make a loop.

2. Take the working end under the loop.

3. Take the working end back over the loop.

Even more secure than
a bowline.

 Difficult to undo.

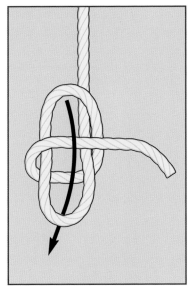

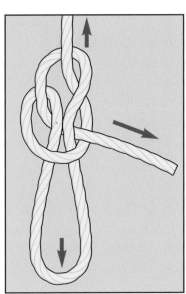

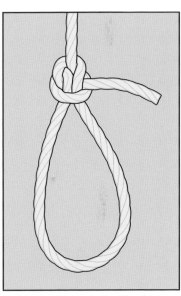

4. Grab the top of the knot.
Take this over both central
parts of the knot and under
the part at the bottom.

5. Pull as shown.

6. The knot now looks like this.

French bowline

Use: To make a double loop, one of which is adjustable.

 Excellent for a body harness, with the adjustable part under the arms.

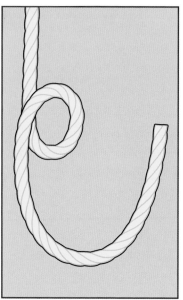

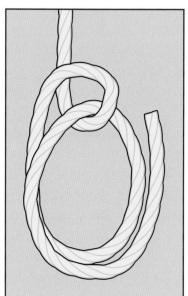

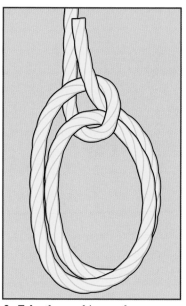

1. Make a small loop.

2. Pass the working end up through the small loop and make a second large loop.

3. Take the working end up through the small loop.

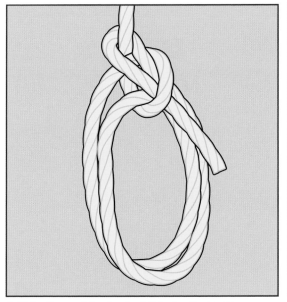

4. Take it round the back of the standing part and down through the small loop.

5. Pull tight.

Stopper knots

A **stopper knot** is used to bulk up the end of a rope, to stop it running through an eye or for throwing.

Summary: We have already learned the figure of eight (left). Here are three more stopper knots, beginning with the overhand knot.

Overhand knot
Use: To stop a rope running out of an eye or block.

 Difficult to undo (a figure of eight is better).

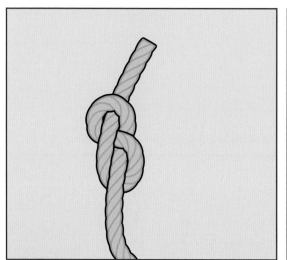

Figure of eight

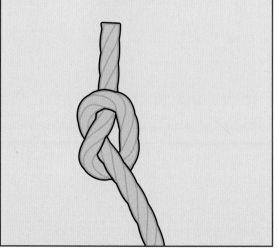

Overhand knot

Stevedore knot

Use: A stopper knot, similar to a figure of eight but with an extra turn.
This gives it more bulk and makes it even easier to undo.

1. Take the working end over the standing part.

2. Take a second turn.

3. Take the working end through the loop.

4. Pull tight.

Heaving line knot

Use: To weight the end of a rope for throwing
(i.e. to use as a heaving line).

1. Make a bight.

2. Now wrap the working end around the bight several times.

3. Pass the end through the loop.

4. Pull the standing part tight to lock the knot.

Monkey's fist

Use: Weighting the end of a rope.

 You can put a ball inside to make the end heavier still. But don't make it too heavy or it might hurt someone.

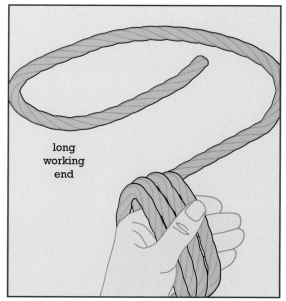

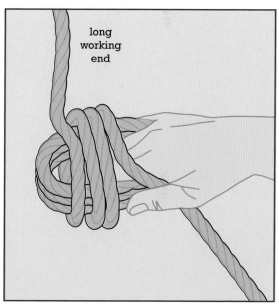

1. Take two metres of the working end. Wrap three equal turns around your hand.

2. Now wrap three turns around the first three, at right angles to them.

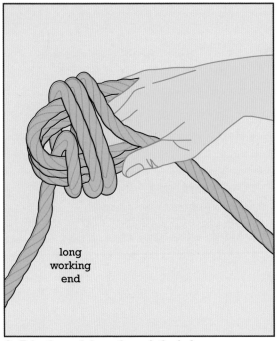

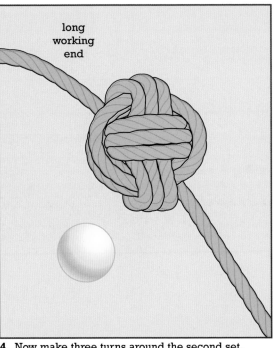

3. Take the end down through the hole.

4. Now make three turns around the second set, going up through the hole on one side and down through the hole on the other side.

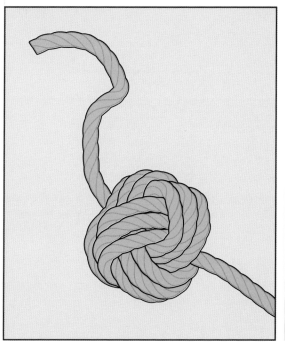

5. (If required, put a weight into the centre at this stage.) Pull tight, starting at one end and working the loose rope through the knot.

6. Finally, cut off the end.

Bindings and lashings

Boa knot

Use: as a temporary whipping or to tie off the top of a sack.

 Very difficult to undo.

 No need to have a free end – can be tied in the middle of a rope.

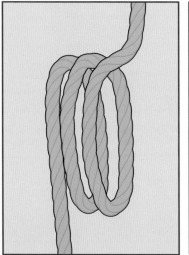

1. Make two loops.

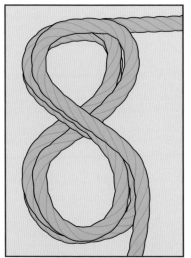

2. Twist the loops. Then pull *up* each side until they touch.

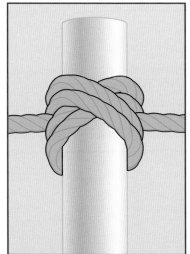

3. Drop the knot over the object and pull tight.

Trucker's hitch / Dolly knot
Use: Tying down.

 Gives a purchase. Can be repeated along the length of an object.

 Pulling the working end undoes the series of knots.

 First loop can topple. Prevent this by using two turns.

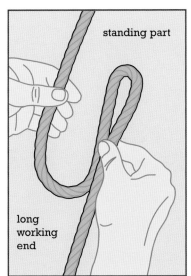

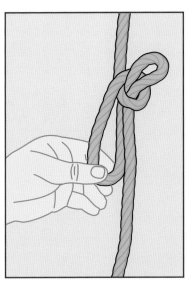

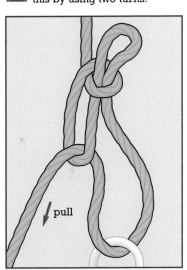

1. To tie down an object to a ring. Make a bight.

2. Put the bight over the standing part and roll. (You can go round twice for safety.)

3. Take the working end through the ring and through the loop. Pull to tension the standing part.

Marling hitches
Use: Lashing a long bundle.

 Must go *over* in step 3, or the lashing will loosen.

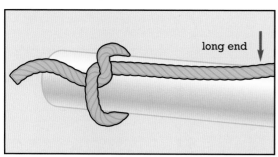

1. The first hitch in place.

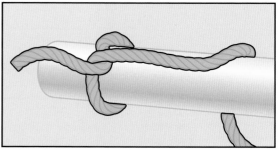

2. Take the working end around the object.

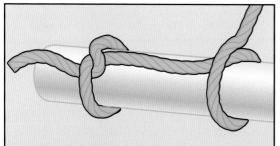

3. Over the standing part.

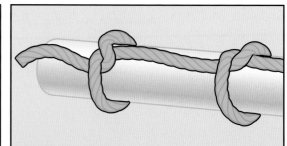

4. Then under. Repeat.

Diagonal lashing

Use: To fix two rods together, at an acute angle.

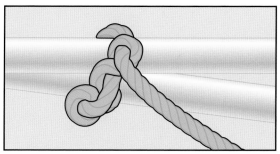

1. Make a timber hitch (see page 35) around both poles.

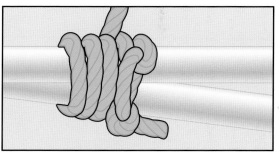

2. Make three turns around the poles.

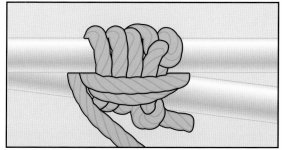

3. Now make two frapping turns (turns at right angles).

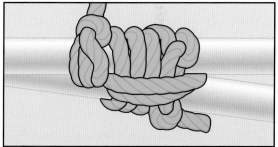

4. Finish off with a clove hitch (page 12) around one pole.

Square lashing

Use: To fix two rods together, at right angles.

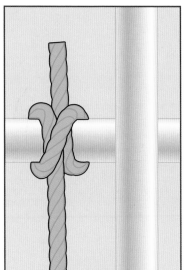

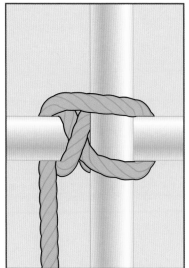

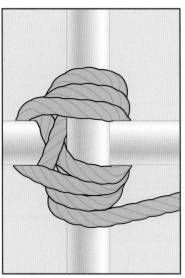

1. Tie a clove hitch on one of the rods (see page 12).

2. Make a turn around both rods: over, under, over and under.

3. Repeat twice. As the rope goes under a rod it should go on the inside of the previous turn; as it goes over a rod it should go on the outside.

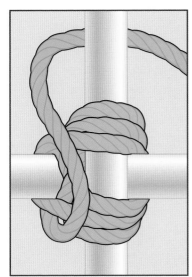

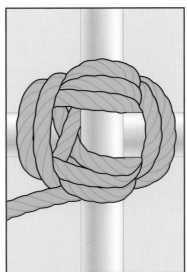

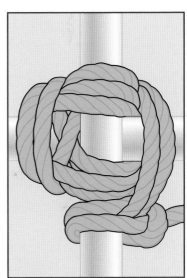

4. Now change the direction of the rope and....

5. ....make two frapping turns (under, over etc).

6. Finish off with a clove hitch.

Bottle knot
Use: A knot which tightens around an object, e.g. the neck of a bottle.

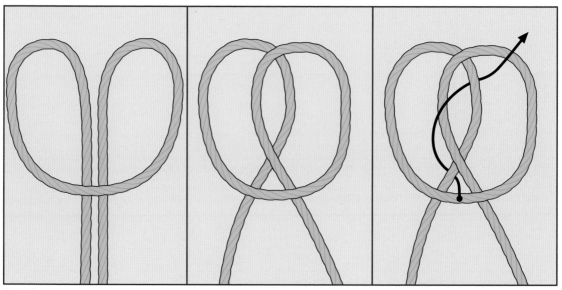

1. Make two loops like this.

2. Place the right-hand loop halfway over the left-hand loop.

3. Take the bottom of the knot under, over, under and over the other parts.

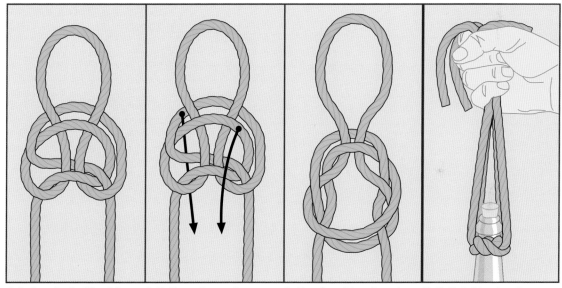

4. It then looks like this.

5. Pull two loops down like this.

6. Put the object into the knot.

7. Pull tight using the two ends and the two parts of the loop.

Shortening a rope

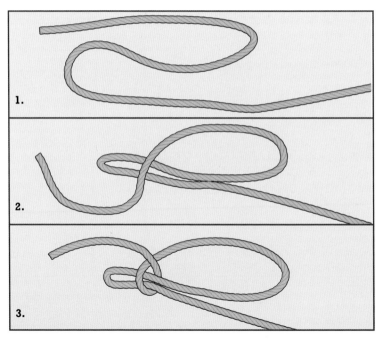

Sheepshank

Use: To shorten a rope temporarily. To protect a piece of chafed rope (chafed bit must be the centre of the 'Z').

 Easily shaken undone.

1. Make a Z in the middle of the rope.

2. Take the left-hand end over.

3. ...underneath, and back over itself.

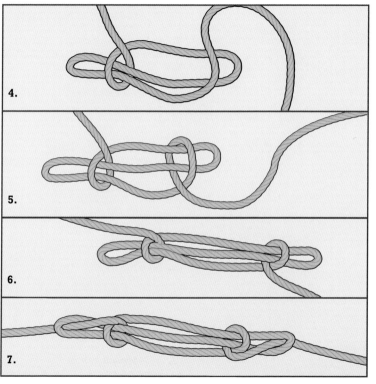

4. Take the right-hand end over...

5. ...underneath, and back over itself.

6. Pull tight.

7. For security, put each working end through its loop.

 Step 7 shows how to make the knot secure, provided the ends are free to go through the loops.

Fishing knots

Grinner knot   **Use:** Tying gut (fishing line) to a hook.

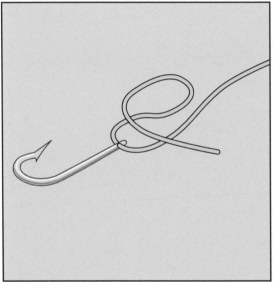

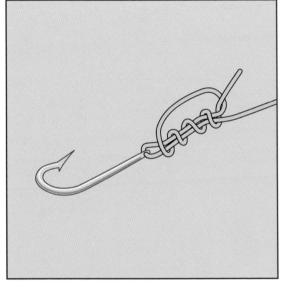

1. Put the gut through the hook and arrange it like this.

2. Take the working end around the standing part and through the loop four times.

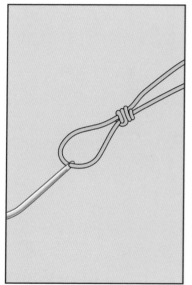

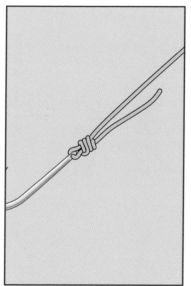

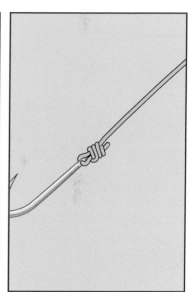

3. Lubricate the knot (with spittle). Pull both ends to tighten up.

4. Slide the knot down tight to the hook.

5. Then cut off the end.

Blood dropper knot

Use: To make a loop in a line, to which you can attach flies, hooks or weights

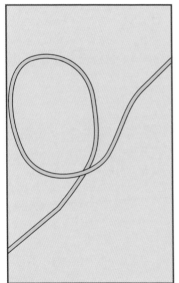

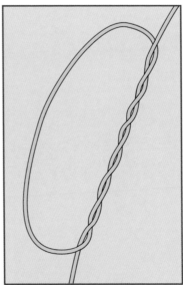

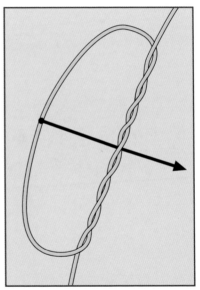

1. Make a loop.

2. Wrap the loop around the standing part at least four times.

3. Find the centre of the twists and pull the loop through it. Lubricate with spittle.

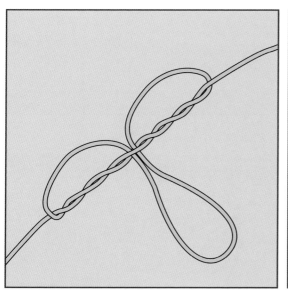

4. It now looks like this.

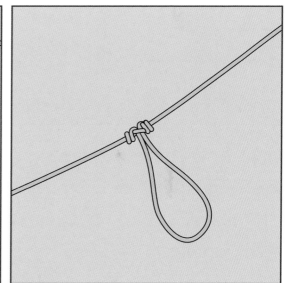

5. Tighten.

Blood knot
Use: Joining two pieces of line (gut).

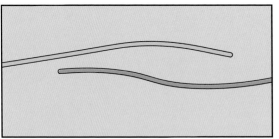

1. Lay the pieces of gut side by side.

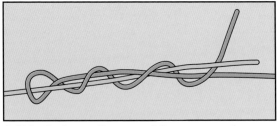

2. Take the brown line around the green one four times, and then in between the two.

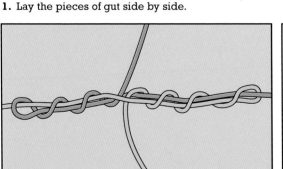

3. Now do the same with the green line.

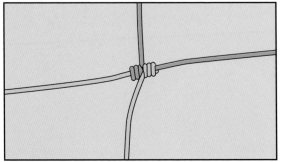

4. Pull tight.

Double fisherman's knot/
Double Englishman's knot

Use: For joining two pieces of line (gut) together, particularly of large diameter.

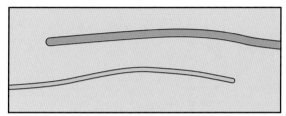

1. Lay the two pieces alongside each other.

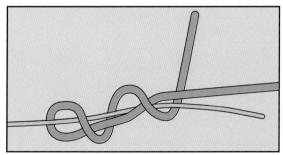

2. Take the end of the brown gut around the green one twice....

3. ....and back through both loops.

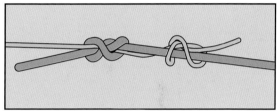

4. Now do the same with the green gut.

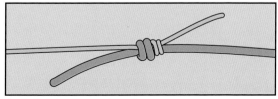

4. Pull tight. Then trim off.

Twisted loop

Use: To make a loop in the middle of a line.

 Simple to tie.

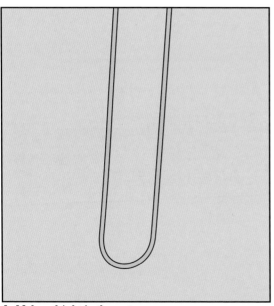

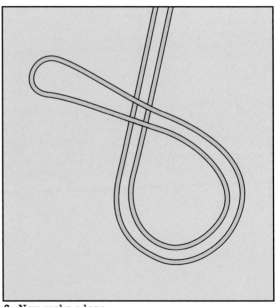

1. Make a bight in the gut.

2. Now make a loop.

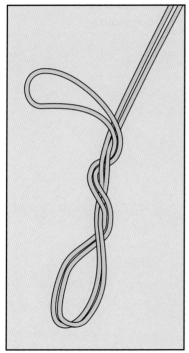

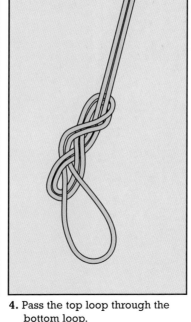

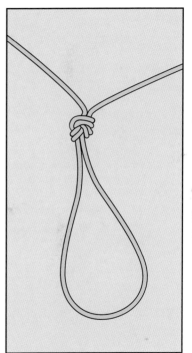

3. Twist the loop at least twice.

4. Pass the top loop through the bottom loop.

5. Pull tight.

Arbor knot

Use: For attaching a fishing line to a reel.

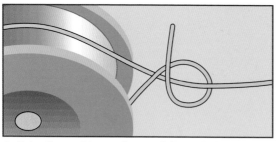

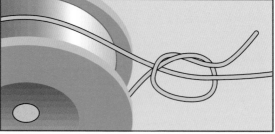

1. Take the working end around the reel, then over and around the standing part.

2. Take the working end back under itself.

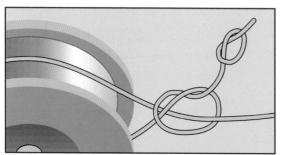

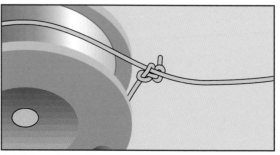

3. Tie an overhand knot in the working end.

4. Tighten by pulling on the standing end. Cut off excess line.

Stop knot

Use: A second piece of gut is used to make a 'lump' in a line to stop a float or weight slipping along it.

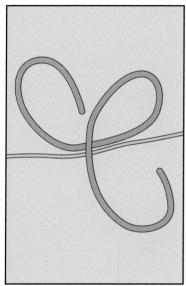

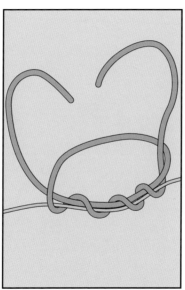

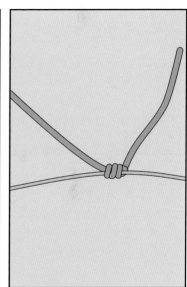

1. Lay the two pieces of gut like this.

2. Take one end of the brown line around the standing part and through the loop, four times.

3. Pull tight, then trim off the ends.

Fancy ropework

Turk's head

Use: Along a tiller (to stop your hand slipping). For decoration.
You can also make it flat for a mat.

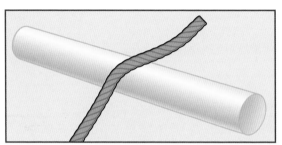

1. Take an end over the object.

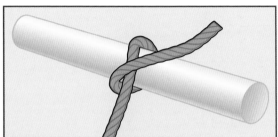

2. Pass it around and over itself.

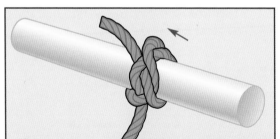

5. Push the right-hand turn over the left-hand turn.

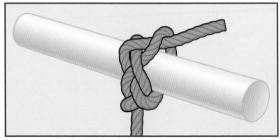

6. Take the end over and under from left to right.

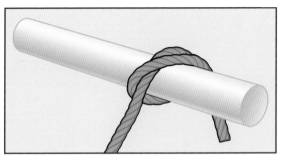

3. Around the object again.

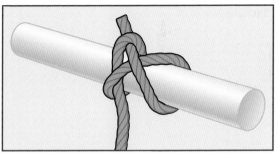

4. Pass the working end over and back under the first turn.

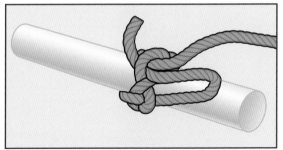

7. Rotate the knot. Pass the end through the first loop, next to the standing part.

You can double or triple the knot by passing the end alongside all the original turns. Here it is doubled.

Portuguese flat sennit

Use: To make a handle so you can pull on a ring, light-switch, snap-shackle etc.

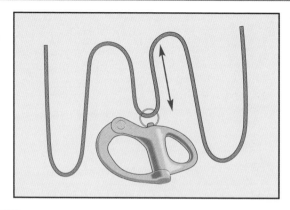

1. Thread the rope through the ring and arrange it like this. The central portion needs to be the length of the final handle.

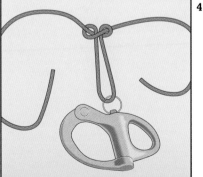

4. Pull fairly tight and it will look like this.

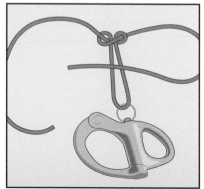

5. Now take the right-hand end over.

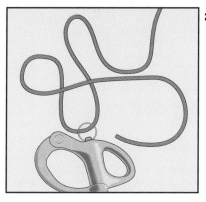

2. Take the left-hand working end over itself, like this.

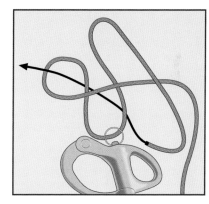

3. Now take the right-hand working end over the left-hand one, under everything, and through the left-hand loop.

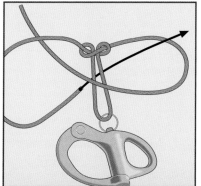

6. The left-hand end goes over, then under everything and through the loop.

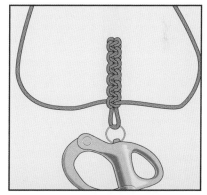

7. Pull tight, then repeat steps 2, 3, 4, 5 and 6 several times. Finally, cut off the ends and heat-seal them.

Ocean mat

Use: To make a decorative mat.

You will need about 4 m of rope.

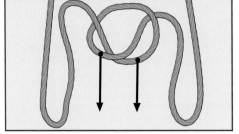

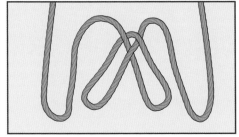

1. Centre the rope and make a loose overhand knot.   **2.** Extend the middle loops downwards (see arrows).

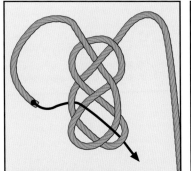

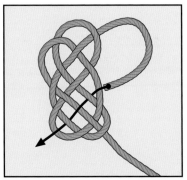

5. Take the left-hand working end under, over, over and under.

6. Pull through so it looks like this.

7. Now take the right-hand end through the mat like this.

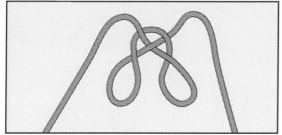

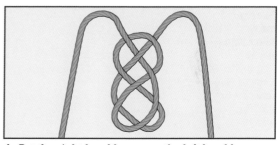

3. Twist each loop clockwise through 180 degrees.

4. Put the right-hand loop over the left-hand loop.

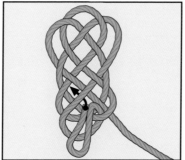

8. Pull through so it looks like this.

9. Continue with the same end. Take it to the bottom right of the mat (where the other end exits).

10. Then take it back through the whole mat, 'following the line'. Do the same with the other end, entering the mat bottom left. Finally, cut off any spare rope and glue the ends out of the way.

Wall and crown knot / Manrope knot

Use: To stop the end of a rope running through an eye.

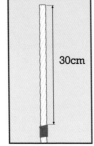

30cm

1. Make a whipping (see page 98) 30 cm (1 foot) from the end.

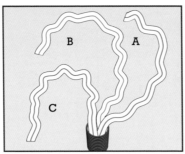

2. Unlay the rope up to the whipping. The ends must be heat-sealed, taped or whipped. Arrange the ends like this.

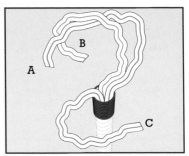

3. Pass strand A under and then over strand B.

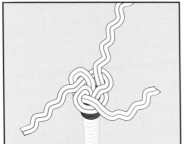

7. Draw the knot reasonably tight. This is a wall knot.

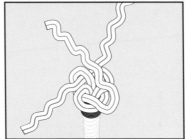

8. Pass one strand across the centre.

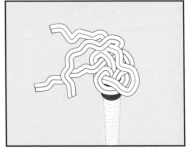

9. Take the next strand (to the right) across the other two.

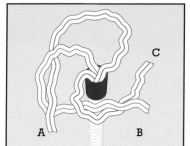

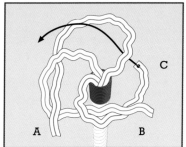

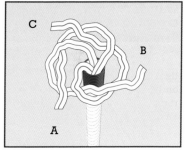

4. Pass C over and then under B.

5. Now pass C around and up through the loop in A…

6. ….like this.

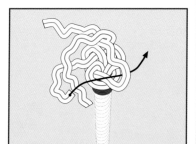

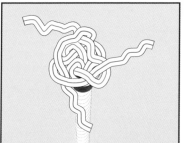

10. Take the last strand over, and through. Pull reasonably tight.

11. This is a wall and crown knot.

12. Choose any strand and take its end through the whole knot, following the strand directly below it to the left or right. Repeat with the other strands. Cut off the ends.

Taking care of a rope's end

Common whipping
Use: Stopping a rope's end unwinding.

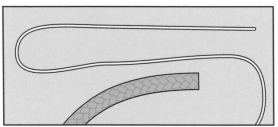

1. Make a bight in the whipping twine.

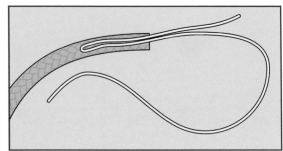

2. Lay the bight along the end of the rope.

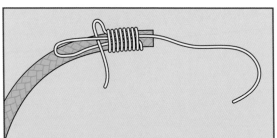

5. Push the end of the whipping twine through the loop.

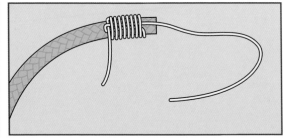

6. Pull the end of the whipping twine until tight.

This is a quick way of whipping. Make it neat and tight or it will not be secure.

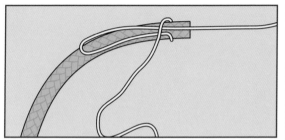

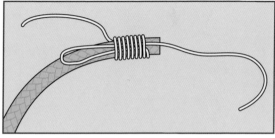

3. Begin to wrap the long end of the whipping twine around the rope and loop, working away from the rope's end. Keep the twine neat and tight.

4. Keep winding until the whipping is 2.5 cm (1 inch) long.

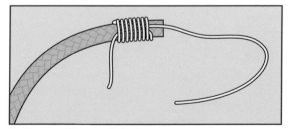

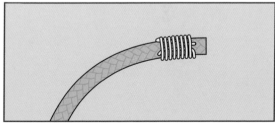

7. Keep pulling until the loop has been pulled halfway under the whipping.

8. Pull very tight. Cut off both loose ends.

Sailmaker's whipping

Use: For three-strand rope.

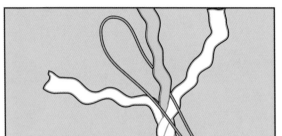

1. Unlay part of the rope. Make a bight in the whipping twine (with a long and short end). Put it over the centre strand.

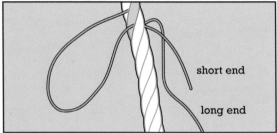

short end

long end

2. Re-lay the rope carefully, to grip the bight.

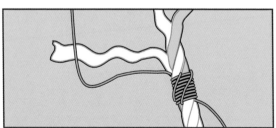

5. Pull tight on the short tail so that the bight clamps tight over the turns of the whipping.

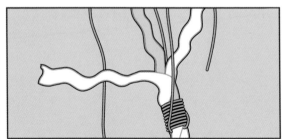

6. Turn the rope over, 180 degrees. Take the short end and with it follow the strand it is beside. Pass the end through the splayed rope.

Hint: To make a sailmaker's whipping on braided ropes use a needle to thread the twine through the rope.

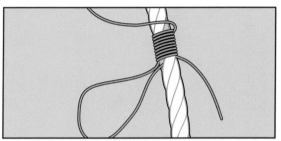

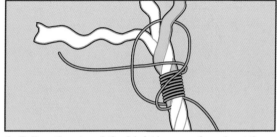

3. Wind the long end of the whipping twine around the rope, against the lay, for 2.5 cm (1 inch).

4. Follow the strand the bight is around, and put the loop over the end of this strand.

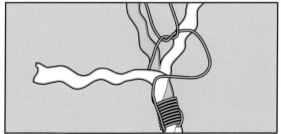

7. Tie a very tight reef knot in the ends of the whipping twine. Check the whipping is symmetrical.

8. Cut off all the loose ends and heat-seal (see p. 102) – except for natural fibres.

Heat-sealing the end of a synthetic rope

(Sometimes called the Butane Backsplice!) **Use:** Heat-sealing can be used as a temporary whipping. But after a while the melted rope cracks. **Objective:** To melt the rope, not burn it!

Needle and palm whipping

 The most secure whipping.

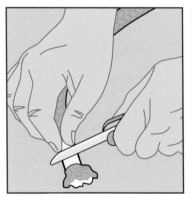

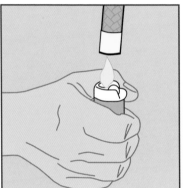

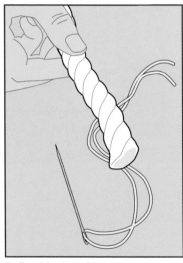

1. Tape the end. Then cut off the end, through the tape.

2. You are going to use a naked flame. Hold the lighter vertical. Light the flame with the rope well clear.

3. Rotate the rope as you lower it towards the flame, so the whole end is warmed. Don't lower too far, or the rope will burn.

 Don't touch the melted rope. Don't let melted rope drip on you.

1. Double the whipping twine through the needle. Push the needle through the rope about 5 cm (2 inches) from the end.

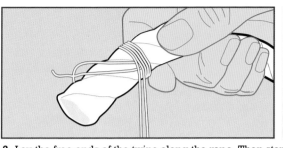

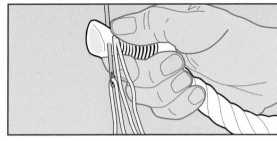

2. Lay the free ends of the twine along the rope. Then start to wind the twine around the rope (and over the ends). After 1.5 cm ($\frac{1}{2}$ inch) cut off the tails. Continue winding towards the end of the rope.

3. When you get near the end of the rope push the needle under one strand.

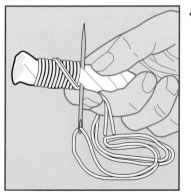

4. Follow the line of the same strand and push the needle under it at the other end of the whipping.

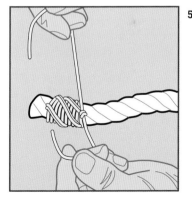

5. Follow back to the other end of the whipping and push under the next strand. Continue until the needle has gone under each strand at each end. Finish off by taking a turn around the last strand and tying a reef knot in the two ends of the whipping twine.

Making a temporary eye or loop

Flat seizing

Use: To make a semi-permanent eye. A seizing is more permanent than a knot but not as permanent as a splice.

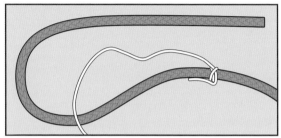

1. Make a loop in the rope. Tie a 2 m length of thin line to the rope with a clove hitch.

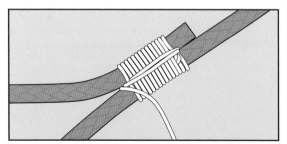

4. Make two frapping turns (wind twice at right angles to the seizing).

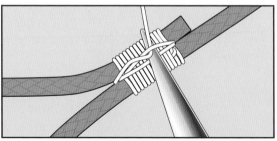

5. Take the thin line under the first frapping turn.

Strength depends on the tightness of the seizing.

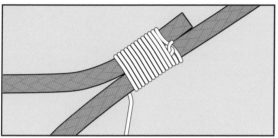

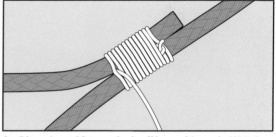

2. Wind the thin line around both pieces of rope, trapping the tail of the clove hitch. Pull tight.

3. After about 12 turns lock off by making a half hitch around one piece of the rope. Pull tight.

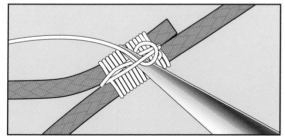

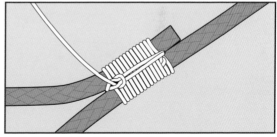

6. Then pass it around the second frapping turn and back underneath it.

7. Pull very tight. Then cut off the end and heat seal it.

Needle and palm seizing

Use: To join two ropes, side-by-side.

 Very secure.

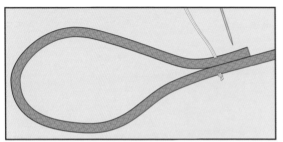

1. Make a loop in the thick rope. Thread the needle with whipping twine, double it up and tie a knot in the end. Push the needle through the thick rope.

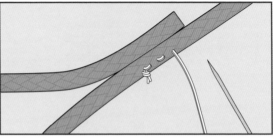

2. Sew through the rope towards the end of the thick rope, leaving 1 cm ($\frac{1}{2}$ inch) between stitches.

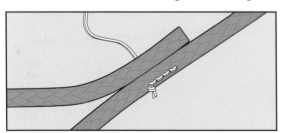

3. Sew back again filling in the gaps.

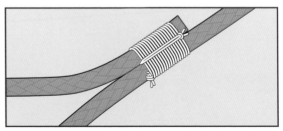

4. Finish off with frapping turns and a knot, as for flat seizing (see pages 104-5).

Joining ropes permanently

Back splice   **Use:** To terminate the end of a rope and make a small loop (for light loads).

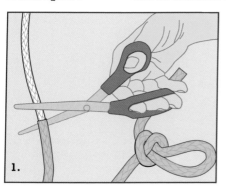

1. Tie a slip knot two metres from the end. Pull back the cover to expose 60 cm (2ft) of core. Cut off this core.

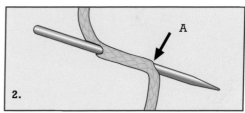

2. Go back to the slip knot and milk the cover back so there is a tail of 'cover only'. About 15 cm (6 inches) from the end of the core, push in a hollow needle. It must exit so it butts up against the end of the core, at A.

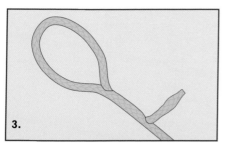

3. Push the end of the cover through the hollow needle. Pull out the needle.

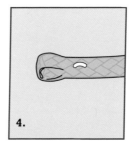

4. Pull the cover through to close up the loop or even further. Then cut off the loose end and rub between your hands to even up. Finally, sew through the overlapped part.

JOINING ROPES PERMANENTLY

108

Eye splice

Use: For making a permanent eye in 3-strand rope.

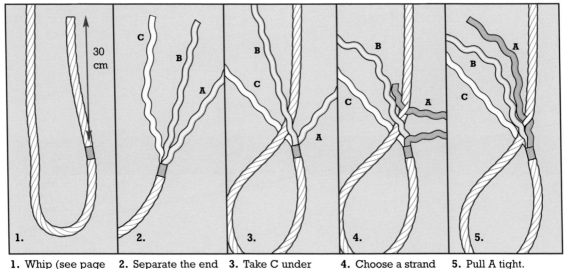

1. Whip (see page 98) the rope as shown. This will stop it unlaying further than is needed.

2. Separate the end into three strands. Whip or tape the ends, or heat-seal them. Note Strand B is on the top.

3. Take C under and B over the standing part.

4. Choose a strand to give the correct size of loop and pass A through it from right to left.

5. Pull A tight.

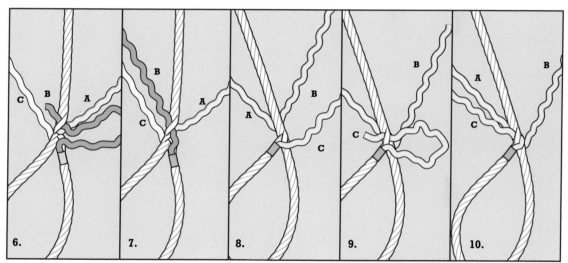

Reduces strength by 30%.

6. Pull A to the right. Pass B under the next strand up.

7. Pull B tight.

8. Now turn the splice over, 180 degrees.

9. Choose the strand that lies uppermost, and is below the two that have been tucked. Pass C under the strand, from right to left.

10. Pull tight. The first tuck is now complete.

Eye splice continued...

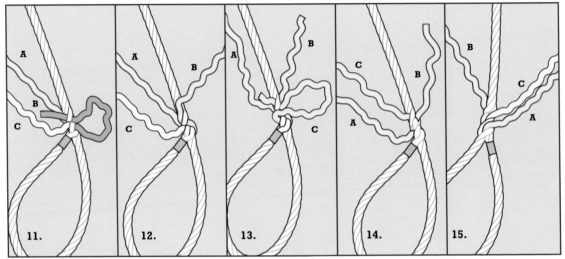

11. The procedure is now to go over one strand and under the next. Start with B and tuck from right to left.

12. Pull tight.

13. Take C over and under, tucking from right to left.

14. Pull tight.

15. Turn the whole splice over (180 degrees).

Reduces the strength by 30%.

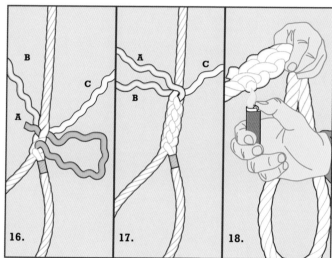

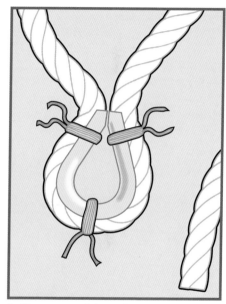

16. Take A over and under, tucking from right to left.

17. Pull tight. Repeat B, C, A, to give five tucks. (For natural fibre only 3 tucks are needed.)

18. Synthetic rope: cut the ends quite close and heat-seal. Natural fibres: cut the ends quite long.

1. For wear resistance incorporate a thimble.
2. Choose one so the rope fits into it snugly.
3. Whip the rope tightly onto the thimble before you begin.

Short splice

Use: To join two pieces of identical 3-strand rope.

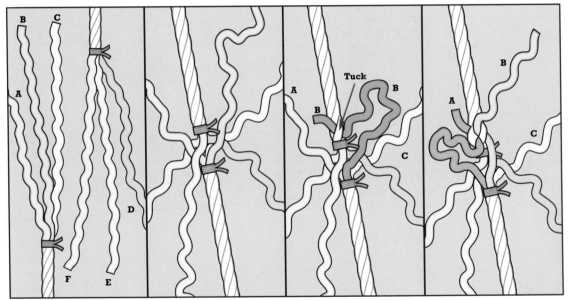

1. Prepare the ends by whipping and unlaying.

2. Interlock the ropes so the strands run alternately.

3. Choose any strand, e.g. B. Make a tuck, from right to left.

4. Now pass A under the next strand, tucking from right to left.

JOINING ROPES PERMANENTLY

The idea is to interlock the ropes, then splice them in a similar
way to the eye splice, with three tucks each side of the join.

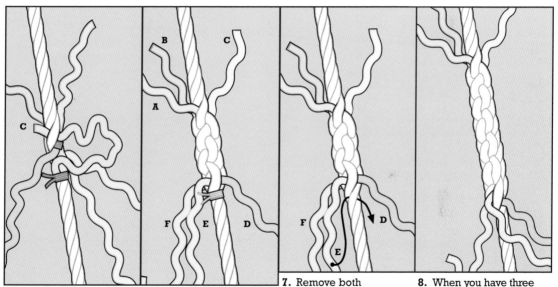

5. Turn the whole splice upside down (180 degrees) and make a tuck with C.

6. Continue tucking (over, under) and rotating until there are three tucks with A, B & C.

7. Remove both temporary whippings. Splice the other side, starting with either D, E or F. (Here it's E.)

8. When you have three tucks on this side too, cut the ends and heat-seal them (synthetic rope). For natural fibres cut the ends quite long.

Eye in braid-on-braid rope

This is similar to the eye on pages 118-119. But the core is braided, not 3-strand. So the cover is pulled down the centre of the core.

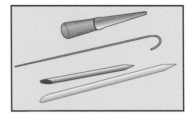

You will need a splicing kit:
a fid, a pusher/needle and two hollow needles.

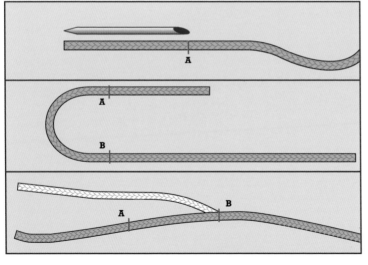

1. Tie a knot two metres from the end to stop the cover slipping. Cut off the heat-sealed end. Make a mark one needle-length from the end – mark A.

2. Make a bight the size of the eye required, and make mark B opposite A.

3. Pull out the core from B.

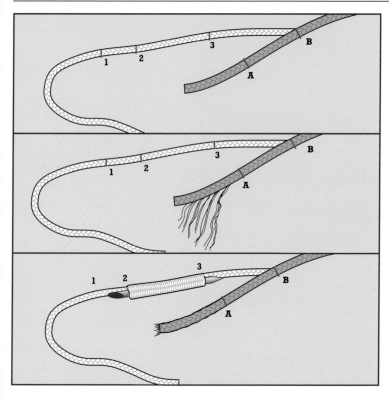

4. Mark the core where it comes out of the cover: mark 1. Pull out more core. One short needle-length from 1, make mark 2. One long needle-length, make mark 3.

5. Reduce the thickness of the cover by pulling out threads and cutting off.

6. Push the needle up the core from mark 2 to mark 3.

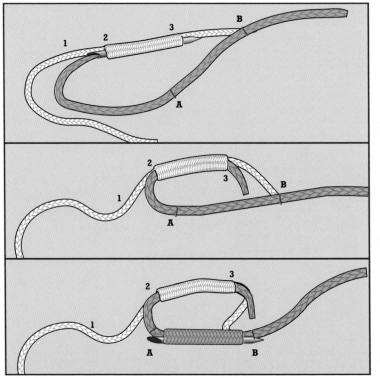

7. Put the (thinned) cover
     end into the needle.

8. Use a pusher to push the cover
     through. Pull out the needle.

9. Now push the needle through
     the cover from mark A to mark B.

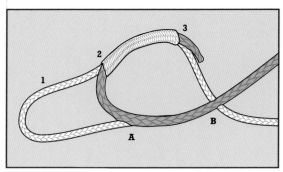

10. Push the core through using a pusher.

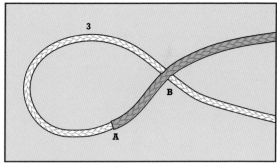

11. Pull both ends until you've smoothed the crossover point.

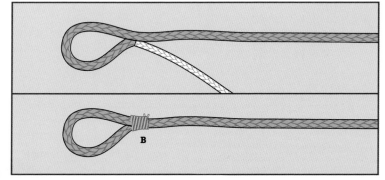

12. 'Milk' the cover back from the knot to the splice. The core will disappear into it.

13. Cut off the loose core at an angle and 'milk' it back in.

14. Put a whipping round point B.

Eye in braid-on-3-strand rope

You are going to make an eye by passing the core, and then the cover, up inside the rope's cover.
You will need your splicing kit (see page 114).

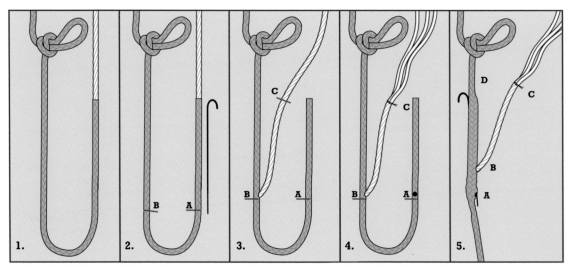

1. Tie a stopper knot 2 metres (6ft) from the end. Pull back the cover to expose 20 cm (8 inches) of core. Milk the slack in the cover down to the stopper knot.

2. Measure one needle-length and mark the cover at A. Form the eye and make a mark, B.

3. Make a hole at B and pull out the core. Whip the core at C so that C B = A B.

4. Unravel the core to the whipping and cut off 50% of each strand. Make a hole at A.

5. Milk the cover. Go back ➤

➤ nearly a needle's-length from B and push a fid into the cover at D. Use it to push the needle through the cover from D, to come out at A.

8. Insert the needle 3 inches below D and out at B (making sure the needle is on the opposite side to the core tail).

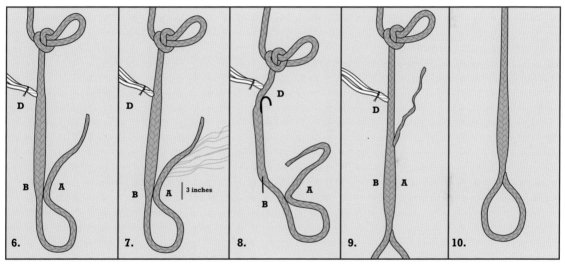

6. Thread the reduced ends of the core through the needle. Withdraw the needle so the core is pulled through the cover and out at D. This forms the eye.

7. Take 7 yarns out of the tail of the cover 3 inches from A and cut them off to make the tail thinner.

9. Pull the thinned cover through the thick cover until the tail comes out. Pull tight.

10. Cut off both tails (core and sheath) and milk until both ends are lost in the cover.

Tapering

Tapering a rope
Use: To taper braid-on-braid rope only.

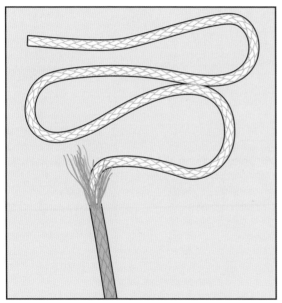

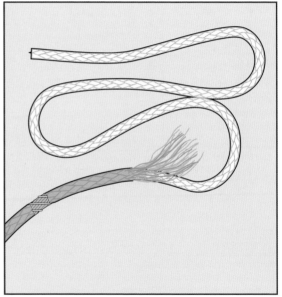

1. Cut off the cover to expose 1m (3 ft) of core.

2. Make a needle and palm whipping about 30 cm (1 ft) from the end of the cover.

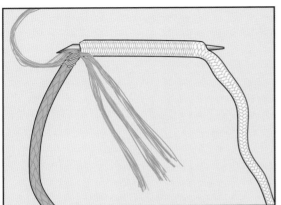

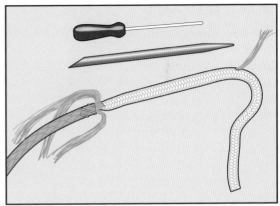

3. Unpick the cover up to whipping. Gather the yarns into four groups. Push a hollow needle through the centre of the core.

4. Use a pusher to push one group of yarns through the core. Then 'milk' the core until the ends of the yarns disappear into it.

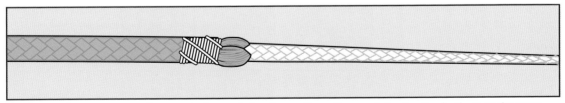

5. Shorten the next group of yarns, and repeat step 4 with it. Shorten the next group still further before threading, and shorten the final group the most. This gives a taper.

Throwing, cleating, winching and coiling

Throwing a rope

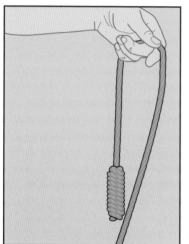

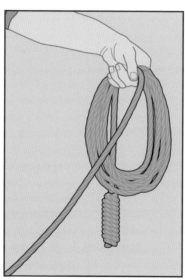

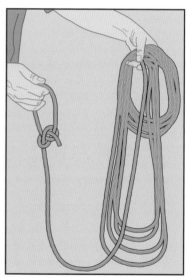

1. Take one end (weighted if necessary) in your left hand.

2. Make about eight small coils (see page 128).

3. Then make larger coils with the rest of the rope. Make a small bowline in the end.

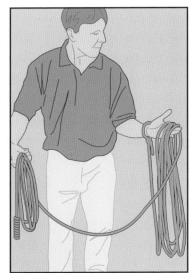 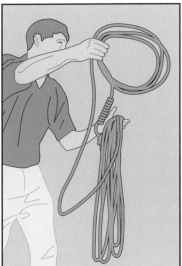 

4. Separate the larger coils and put them in your left hand with the loop over your wrist. Take the small coils in your right hand.

5. Throw the small coils.

6. Let them pull the large coils off your open left hand.

Cleating a rope

Use: Securing a rope temporarily.

1. Bring the rope to the cleat.
2. Make a dry turn (once around).
3. Begin to make a figure of eight....
4. ....and complete it.
5. Make another figure of eight.
6. Make another dry turn.
7. To finish off, coil the rope (see page 128).
8. Put your hand through the coil.
9. Pull back a loop.
10. Twist it twice.
11. Hang the twisted loop over the cleat.
12. There you go – secure, neat and tidy.

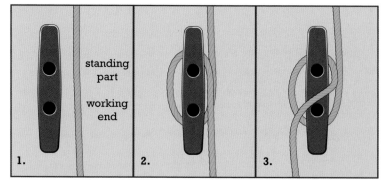

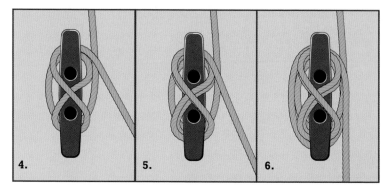

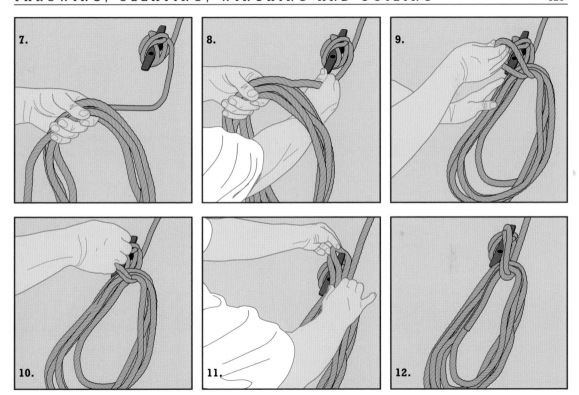

Winching

This is a self-tailing winch.

Hint: Always take a rope clockwise round a winch.

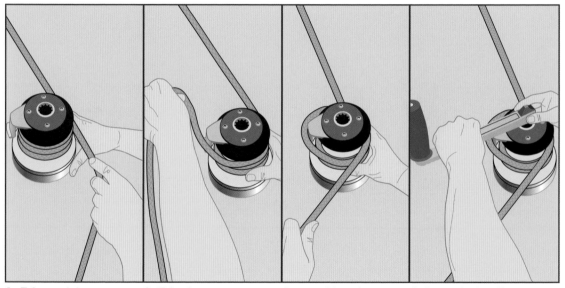

1. Take a minimum of three turns around the winch.

2. Take the rope over the arm....

3. ....and into the grip at the top.

4. Insert the handle.

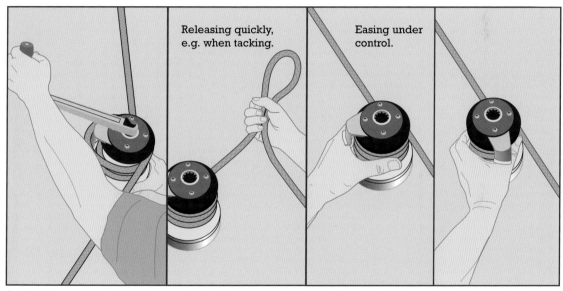

Releasing quickly, e.g. when tacking.

Easing under control.

5. Wind!

6. To release quickly, take out the handle. Then take the rope out of the grip with your right hand.

7. To ease slowly use your right hand to keep tension on the rope. Put your left hand around the drum...

8. ...and carefully ease out the rope with both hands.

Coiling a rope

The secret is to twist and coil the rope in a clockwise direction to stop it kinking. Finish with two or more turns and push the end through the loop.

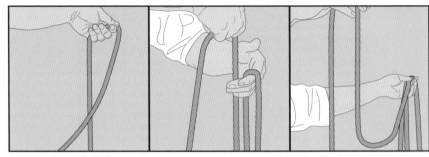

1. Take the end of the rope in your left hand.

2. Twist the rope clockwise in your right hand and transfer the rope to your left hand.

3. Stretch the rope out with your right hand.

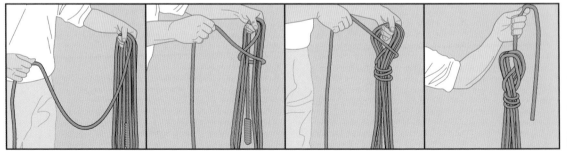

4. Repeat steps 2 and 3 several times.

5. After you've coiled the rope take the working end around the coils.

6. Make several turns.

7. Put the working end through the loop and pull.